Fresh Cow Problems

HOW TO CONTROL THEM

HOARD'S DAIRYMAN

Contents

Introduction 4
Dry cow care & feeding 5
    Drying off; length of dry period; body condition
       score; forages; fiber; concentrates; water;
Table of nutrient requirements for dry cows 7
Body condition score photos 8
Udder edema 10
Milk fever 11
Ketosis 15
Fat cow syndrome 17
Retained placenta 18
Displaced abomasum 20
Rumen acidosis 22
Laminitis 23

Copyright © 1993 by W.D. Hoard & Sons Company

All rights reserved. No part of this book may be reproduced or used in any form or by any means, electronic or mechanical, including photocopying, recording or by any information storage and retrieval system, without permission in writing from the publisher.
Address inquiries to W.D. Hoard & Sons Company,
28 Milwaukee Avenue West, Fort Atkinson WI 53538.

ISBN 0-932147-16-X

Printed in the United States of America

Foreword

We are proud to present this second edition of FRESH COW PROBLEMS. Since first published in 1986, over 20,000 copies have been sold. Demand actually has increased over the years, as has the awareness that the dry cow requires the same level of management as her milking herdmates, if she is to avoid costly health problems upon freshening and if she is to produce up to her potential in her upcoming lactation.

The backbone of this book is the writing of L.H. "Bud" Schultz, now retired from teaching and research in dairy science at the University of Wisconsin. His research focused on metabolic disorders and mastitis, and he authored the original sections of the book.

Drs. Lee Allenstein and Gary Oetzel, faculty associate and assistant professor with the University of Wisconsin's School of Veterinary Medicine, reviewed and added new material to this edition. Dr. Allenstein is a retired veterinary practitioner and columnist for Hoard's Dairyman magazine. Dr. Oetzel's valuable updating of the section on milk fever is based on his research on calcium metabolism and feeding anionic salts to prevent milk fever. He also added a new section on rumen acidosis. Dr. Allenstein wrote a new section on laminitis.

We thank all three authors for focusing on this important area of dairy management. Our hope is that this updated and expanded edition will add to the value of our dairy library of books, a service to our readers.

HOARD'S DAIRYMAN

Introduction

Calving and the first month after freshening are critical times for the dairy cow. Most problems occur at these times. Ask any veterinarian.

In addition to calving difficulty, there can be retained placenta, udder edema, milk fever, displaced abomasum, ketosis and fatty-liver syndrome. These are metabolic rather than infectious diseases and are related primarily to feeding and management. Some infectious problems, such as mastitis and metritis, are also more prevalent at these times.

Metabolic diseases are completely interrelated and tend to occur together. For example, if a cow develops milk fever, she is four times more likely to retain her placenta. A retained placenta increases the likelihood of ketosis by more than 16 times. These relationships underscore the need for disease prevention through better overall dry cow management and nutrition.

During the dry period, the pregnant cow's nutritional needs to maintain herself and provide for a growing fetus appear small compared to the needs of producing milk. Problems with dry cow nutrition are not immediately apparent, since these cows are not contributing to the herd's milk production. Dry cows are therefore often relegated to a single group of less importance than that of lactating animals.

Yet evidence grows showing that greater attention to dry cow care and nutrition can pay big dividends in the following lactation. The dry period should be regarded as a time critical for lactation preparations, rather than a mere rest period between lactations.

The greater the cow's genetic ability to produce milk, the greater the adjustment. Beef cows with low milk production have little or no udder edema, milk fever or ketosis. As we select for higher milk production, we need to improve feeding and care to prevent problems. We can't eliminate these problems, but we can keep their incidence at low levels.

This discussion will stress your role in the control of metabolic disorders through feeding and cow care. Emphasis is placed on occurrence, symptoms, causes and prevention. There will be brief discussions of treatment, but your veterinarian should play a major role in this area.

Dry cow care & feeding

A good procedure for **drying off** cows is to discontinue all grain feeding, offer only poor-quality forage and in those producing less than 22 to 24 pounds of milk per day, stop milking abruptly. It may take up to a week to bring production down to this level. Often it helps if cows can be separated from their milking herdmates. Don't allow them to go through the parlor. Keep them out of the milking barn.

An occasional cow may have to be milked once a day for a few days to bring her production down to 20 pounds. Whenever she is milked, she should be milked out completely.

When milking is ceased, dry treat all quarters immediately, and teat dip. If possible, continue teat dipping for up to a week.

The **length of dry period** is ideally 50 to 69 days. Fewer than 40 days reduces subsequent milk yields; over 70 days results in reduced milk yields or in increases that don't make up for the added days the cow is dry.

The dry period is a good time for trimming feet, worming and other parasite control, and vaccinations that are safe during pregnancy. Vaccines to promote needed antibodies in the colostrum also can be considered at this time.

The **body condition score** of cows should not change appreciably during the dry period. Ideally, cows should calve with the same body condition score they had at drying off. Therefore, the most important time for body condition scoring is about 150 days into lactation. This is the key time to prepare for how a cow will score at drying off. It's easiest and least risky at this point to adjust the cow's condition, if needed.

The goal is to have a cow in "good" condition at calving: not too thin and not too fat. The accompanying body condition photos show cows rated 1 through 5.

Few cows would fall in the very thin category. They obviously are not prepared to start milking again. Cows scoring 1 and 2 definitely will produce less during the following lactation because they do not have adequate reserves to use in early lactation.

Generally, cows peak in milk at 4 to 6 weeks after calving, but feed intake does not peak until 8 to 12 weeks. So good cows must draw on body reserves. The better the appetite, the sooner the cow will peak in feed intake and the higher she will peak in production.

Although the very fat cow has lots of reserves to produce the best theoretically, her feed intake lags and she has more metabolic problems, so milk production suffers.

Cows rated 4 produce about 400 pounds more fat-corrected milk than cows rated 3. Part of this advantage likely is a higher milk fat test. The more fat a cow burns off in early lactation, the higher the milk fat test. But you are walking a thin line, because cows with a rating of 5 produce about 1,300 pounds *less* fat-corrected milk the next lactation than cows rated 3.

Excellent feeding management is required to obtain optimal milk production from cows that calve with a score of 4 or higher.

The safest goal is a "3" rating with some leeway on the high side. If you can get the fat cow through calving and the first month or so without problems, she may produce more milk and have a higher test. But it's not worth the risk.

Rather than try to adjust body condition during the dry period, it is better to feed cows so they are in the right condition at drying off. Then hold them there during the dry period. There are several reasons for this. One is that the cow uses her feed more efficiently while milking than when dry, so it is easier and less expensive to put weight on a thin cow. The extra grain also may result in more milk in late lactation.

If a cow is thin at drying off, a 2-month dry period usually is not enough time to put on proper condition. Cows typically can gain only one-fourth to one-half unit of body condition score during the dry period.

On the other extreme, if the cow is too fat at drying off, it usually is not possible to take that fat off and back her down to the proper condition during the dry period. Reversing the process by starvation creates potentially disastrous problems with possible fat deposition in the liver and severe ketosis in early lactation.

Forages should constitute a major portion of the dry cow ration. Forages are the least expensive source of the nutrients needed by dry cows, plus they promote good rumen health.

Legume forages typically contain excessive amounts of calcium for dry cows, unless anionic salts are fed. Calcium and potassium content may limit legumes to 30 to 50 percent of the diet dry matter. Corn silage can make up 30 to 40 percent. A higher percent will provide excess energy as well as lack sufficient protein. Combining these forages with cornstalks, straw or grass forage can dilute excess nutrients to levels appropriate for dry cows. All forages should be analyzed to properly balance the dry cow's nutrient requirements.

Forages won't have as great a benefit to rumen health unless they contain adequate effective *fiber*. Try to avoid finely chopped ensiled forages, or if they must be fed, be sure to include long, coarse dry hay in the ration. A total mixed ration will prevent the dry cow from choosing between a bunk and hay feeder and will ensure sufficient intake of all nutrients.

Concentrates have a place later in the dry period where a "steam-up" period can ease transition to the lactation diet. Increasing grain up to 8 to 12 lbs. per day, and 14 to 16 percent crude protein helps rumen papillae and microorganisms prepare for the upcoming high-grain lactation diet. Providing an adequate transition diet is the most important aspect of dry cow feeding. Cows cannot be managed to prevent metabolic diseases if they are fed in one group for the entire dry period.

Using current lactation concentrate mixes can be risky, since they contain high levels of sodium and potassium that the dry cow should not receive.

Pregnancy increases the cow's *water* intake by over one-third. Clean, good quality water should always be available to the dry cow.

Incorporating all mineral supplements into the ration is a much preferred method to ensure intake over a free choice mineral program.

National Research Council (NRC) recommendations give the average daily requirement for the dry period. This may result in cows being overfed energy in the early dry period and underfed energy in the last 5 to 6 weeks compared to actual requirements.

Managing dry cows in a two-tier system (far-off dry cows and transition dry cows) best matches increasing pregnancy requirements and declining intake ability. The early dry cow needs high fiber/low energy density, while the close-up cow needs higher energy density with less fiber. A two-group system results in less metabolic disease, improved dry matter intake after calving, and increased peak milk production.

Recommendations also don't consider environmental conditions. Severe cold will require increased energy to maintain body temperature, and this is compounded when animals are wet and muddy. Maintenance energy requirements can be dramatically increased by the cow's level of activity and adverse weather conditions.

Minimum nutrient requirements (dry matter basis) for dry cows

Nutrient	Far-Off Dry Cows	Transition Dry Cows
Dry matter intake, % body weight	1.8 - 2.0%	1.6 -1.8%
NEL,[1] Mcal/lb	.58	.68
Crude fat (CF), % (maximum)	5	6
Crude protein (CP), %	13	15.5
UIP,[2] % of CP	25	32
Acid detergent fiber (ADF), %	30	24
Neutral detergent fiber (NDF), %	40	32
Forage NDF, %	30	24
NFC,[3] % (maximum)	32	38
Calcium, %	.50	.50
Calcium, % (range)	—	<.50% or >1.30%
Phosphorus, %	.25	.30
Ca: P Ratio	1.5 to 5.0 :1	
Magnesium, %	.20	.25
Potassium, %	.65	.65
Sulfur, %	.16	.20
Sodium, %	.10	.10
Chlorine, %	.20	.20
DCAD,[4] meq/kg	—	< 0
Cobalt, ppm	.10	.10
Copper, ppm	12	15
Iodine, ppm	.60	.70
Iron, ppm	50	60
Manganese, ppm	40	50
Selenium	.30	.30
Zinc, ppm	50	60
Vitamin A, KIU/lb	1.8	2.2
Vitamin D, KIU/lb	.75	1.0
Vitamin E, IU/lb	12	15 - 40

1 NEL = Net energy for lactation.
2 UIP = Undegradable intake protein.
3 NFC = Non-fiber carbohydrate, calculated as 100 - CF - CP - NDF - Ash.
4 DCAD = Dietary cation-anion difference, calculated as milliequivalents (Na + K) - (Cl + S); lower DCAD values are associated with lower incidence rates of milk fever.

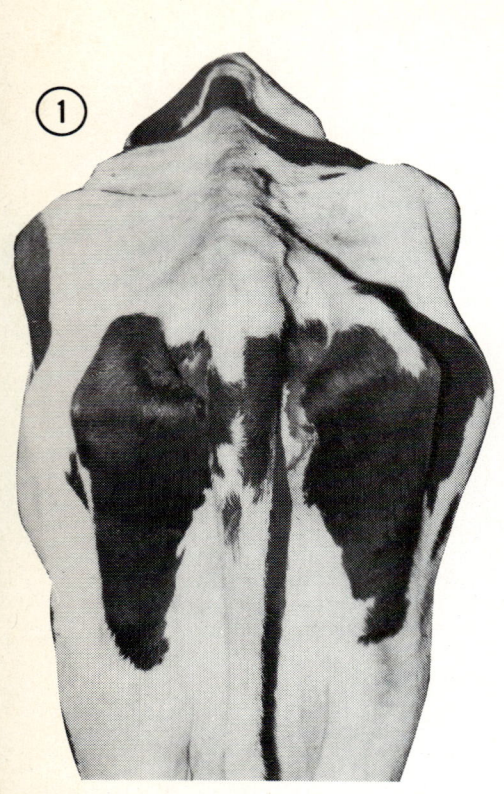
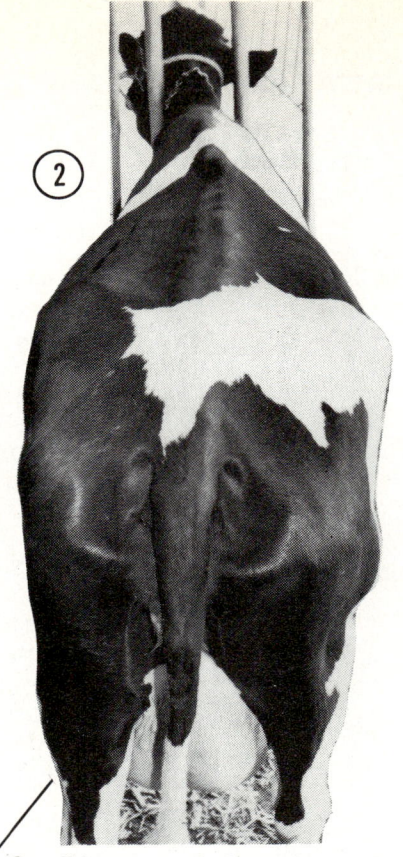
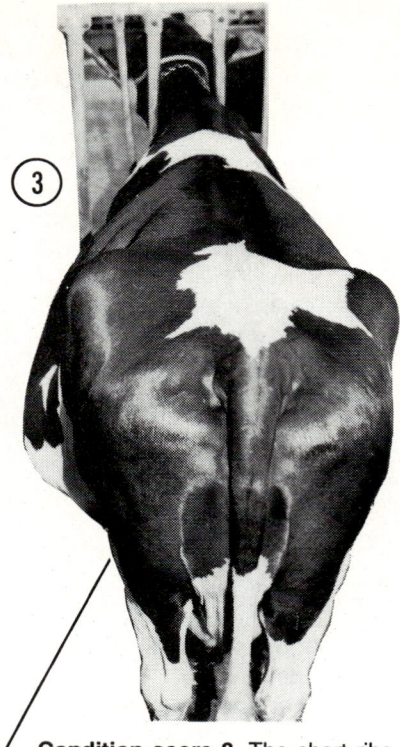

Condition score 1. Individual vertebrae are distinct along the topline; ends of short ribs are prominent, and give a distinctly shelf-like appearance. Hips and pin bones are sharp; severe depressions show between hips and pins, and between hooks. A deep "V"-shaped cavity is below the tailhead and between the pin bones.

Condition score 2. Short ribs can be seen, and the "shelf" still is visible. While hips and pins still are prominent, individual vertebrae along the topline are not distinct to the eye. The depression around the tailhead and pin bones is more "U"-shaped.

Condition score 3. The short ribs appear smooth, with no shelf effect. The backbone is a rounded ridge; individual vertebrae are not visible. Hip and pin bones are rounded and smooth; the area between pin bones and tailhead has smoothed out.

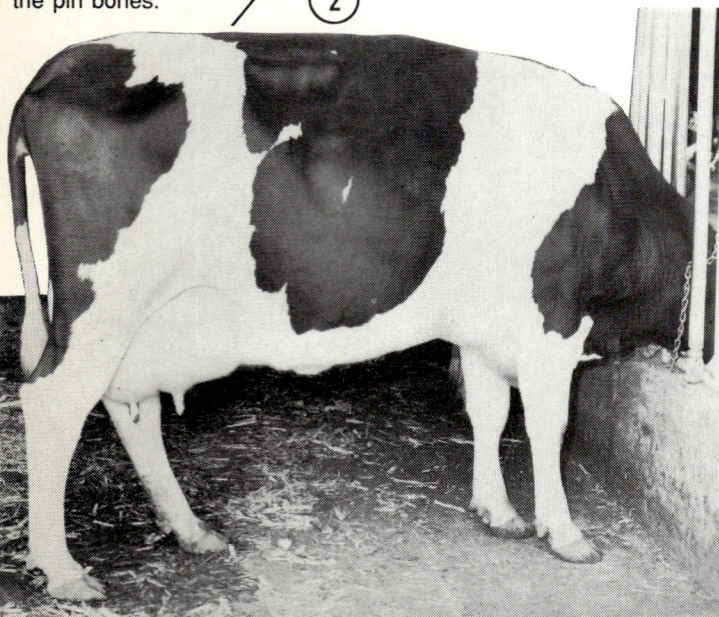

8

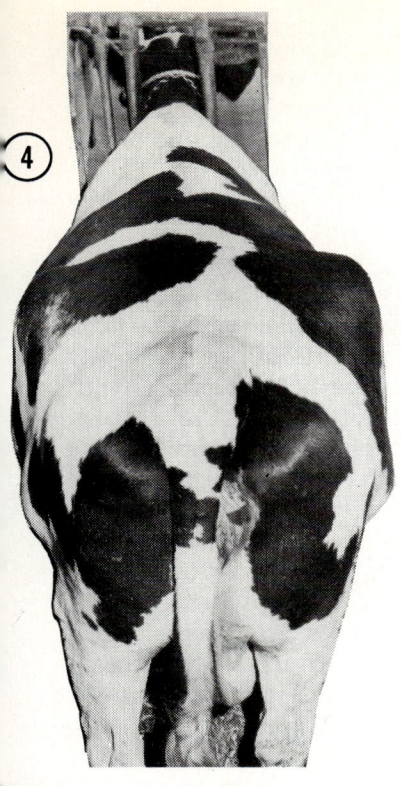
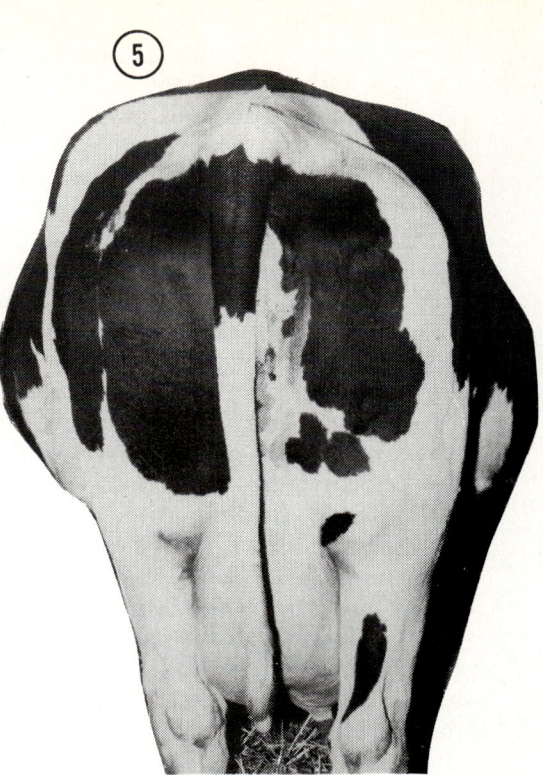

Condition score 4. Short ribs are smooth and rounded; no shelf effect. The chine region is rounded and smooth; loin and rump appear flat. Hips are rounded and flat in between. There is a rounded appearance over the tailhead and pin bone area, with evidence of fat deposit.

Condition score 5. The backbone is covered by a thick layer of fat, as are the short ribs. Hips and pins are not apparent, and the areas between them are rounded; the tailhead is surrounded by fat.

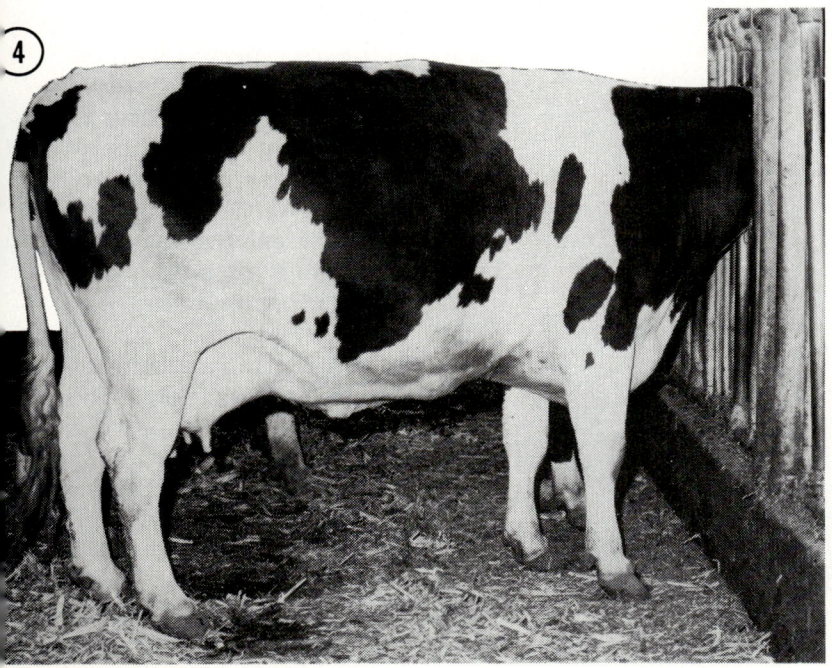

Udder edema

Occurrence. Udder edema usually occurs at calving, sometimes becoming serious before calving. Both heifers and older cows are affected. A cow that has had edema is more likely to have it again.

A Canadian survey of 12,000 cows showed an incidence of 18 percent, but only 1 percent received veterinary treatment. Another study showed a 43 percent incidence, with 5 percent severely afflicted.

Symptoms. There is excessive accumulation of fluid between the secretory cells of the udder, causing swelling. Often, this fluid extends forward under the skin in front of the udder. The source of the fluid is blood serum passing through capillaries to the interstitial tissues.

Gravity causes most of the accumulation to occur in the bottom of the udder between the skin and secretory tissue. The fluid is clear and held rather firmly by a network of connective tissue.

Although edema seems to have no direct adverse effect on production, the indirect effects are undesirable. The udder is sore, and this may interfere with milk let-down. Teats are shortened and tend to strut outward, so the milking machine may not stay on well. There can be physical damage, such as cracking of the udder or harm to attachments in severe cases.

Cause. We don't know the exact cause of udder edema. Measurements of the levels of blood components, such as protein, sodium and potassium, or the osmotic pressure of the blood do not show a significant relationship to edema. Blood pressure changes and impaired lymph flow out of the udder may be involved.

Kentucky workers have shown that very high salt intake before calving (0.5 pound per day) increased the severity of udder edema. They fed 15 pounds of grain per day for 30 days before calving with and without about 3.3 percent salt. Edema was more severe in cows receiving the added salt. This suggests that the salt was not the basic cause of the problem, but aggravates it.

Feeding sodium bicarbonate to dry cows increases their intake of sodium, and increases clinical signs of udder edema. However, feeding anionic salts to prevent milk fever also may help to prevent udder edema.

Another mineral interaction that contributes to udder edema is excessive intake of potassium. This most often is due to overfertilized forages (potassium content greater than 3.0 percent).

There is a common belief among dairymen that heavy grain feeding (over 12 lbs./day) during the dry period causes more edema. However, many of the experiments attempting to produce edema with heavy grain feeding have been unsuccessful. Although heavy grain feeding before calving may not cause edema, it is not recommended for other reasons.

Treatment. Certain mechanical procedures may help. The lymph vessels exit out of the upper rear of the udder. They contain one-way valves which help the fluid move upward.

The fluid is moved forward along the major lymph vessels by suction created by breathing. Ultimately, fluid is dumped

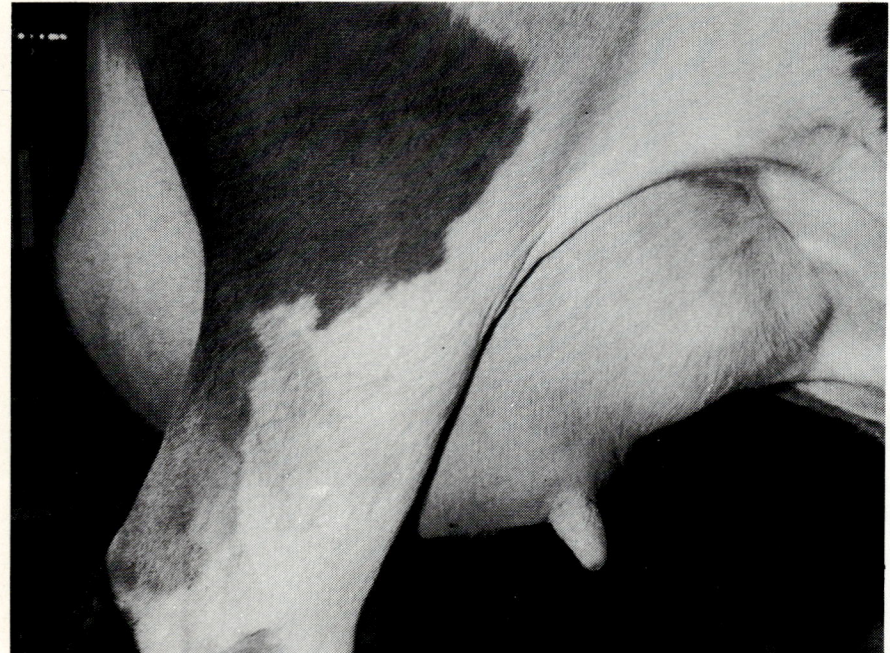

Massage and moderate exercise can help reduce edema. If diuretic drugs are used, milk must be discarded for at least 48 hours.

back into the bloodstream near the heart. Therefore, massaging the udder in an upward direction for 10 to 20 minutes at least twice a day after milking will help move the fluid out of the udder. Fluid returns after the procedure is stopped but to a lesser degree each time.

Udder supports may be helpful for cows with poorly attached udders. Moderate exercise stimulates lymph circulation. Complete milk-out helps resolve udder edema but may be difficult to accomplish when the udder is swollen and painful. Oxytocin injections may aid in milk let-down.

The use of diuretic drugs, which speed up removal of water from the body, may help shorten the period of edema. These sometimes are combined with corticosteroids which have an anti-inflammatory action.

Corticosteroids are potent and should be used cautiously under veterinary supervision. They usually cause a temporary drop in milk production. Milk needs to be discarded for at least 48 hours following administration of diuretics. Follow label instructions.

Prevention. Avoid feeding excess salt during the dry period. The salt requirement of dry cows is about 0.25 percent of the total diet. One ounce per day (28 grams) should meet the requirement easily, as will a modest level of grain feeding with 1 percent trace mineral salt. If no grain is fed, self-feeding of salt is adequate.

Provide moderate exercise along with good care before calving; prevent chilling, bruising and irritation of the udder.

In severe cases, starting milking about 10 days before calving, combined with regular massage, may help. Remember that the colostrum is removed when you start milking, so you need to freeze it to have colostrum available for the calf. After calving, regular massage after milking should speed the udder's return to normal.

Heritability of edema is low, so selecting against it will be a slow process. But culling cows with severe edema should help.

Milk fever

Occurrence. The incidence of milk fever has risen in recent years. The statistical reporting service in Wisconsin suggests that about 8 percent of the cows are affected and about two-thirds of the herds have the problem. In a 1990 survey, more than 82 percent of Hoard's Dairyman readers reported milk fever in their herds, with over 8 percent of the cows affected. In cows beyond the first lactation, incidence probably would exceed 10 percent.

Reports indicate breed differences, with Jerseys having a higher incidence. This suggests that the tendency toward milk fever might be inherited, but most genetic studies reveal that heritability is quite low.

It is doubtful that the greater incidence today is due to genetic selection. Rather, increases probably are due to changes in the way we feed and handle our cows, and the most progress in controlling milk fever can be made through improvements in these areas.

Incidence of milk fever is related to age. It rarely occurs among first-calf heifers and seldom at second calving. There is a progressively higher incidence with each freshening. Cows having milk fever once are apt to repeat. Canadian studies reported that about half of the field cases were in cows with a previous history of milk fever.

The timing of the problem in relation to calving also is unique. Canadian studies reported that 75 percent of cases occurred between 1 and 24 hours after calving. Only 3 percent occurred before calving, 6 percent at calving, 12 percent between 25 and 48 hours after calving and 4 percent beyond two days.

Symptoms. The name "milk fever" is a misnomer because the cow does not have a fever. Calving paralysis probably would be a better name. There is a lack of appetite and an inactive digestive tract. Often the cow is dull and listless with cold ears and a dry muzzle.

The first specific symptom is incoordination when walking. This progresses to where the cow may fall or lie down and be unable to rise. Canadian workers divided the progress into three stages: I. Standing but wobbly; II. Down on chest and drowsy and III. Down on side and unresponsive.

The major change in the blood of milk fever cows is blood calcium. Normal level is 8 to 10 milligrams (mg.) per 100 milliliters (ml.). Normal cows drop to about 8 at calving. Milk fever cows drop to about 6.5, 5.5 and 4.5 mg./100 ml. in stages I, II and III. It is likely that the drop in blood calcium, accompanied by a drop in phosphorus and an increase in magnesium levels, brings on the symptoms.

Cause. There is reasonable agreement that the cause of milk fever is the extra calcium drain from the blood into the milk at calving. This is coupled with the inability of the cow to change her metabolism rapidly enough to keep blood calcium levels up. She has some built-in mechanisms to do this, but they sometimes do not work fast enough. The way she is fed during the dry period influences the speed of response.

There are two ways the cow can get more calcium — mobilize it from bone or absorb it from the digestive tract. The two important compounds in the body which influence these processes are parathyroid hormone (PTH) and a substance we'll call active vitamin D.

PTH comes from the parathyroid gland located in the cow's neck. Its release is triggered by low blood calcium. The hormone's major effect is to cause calcium to move from the bone into the blood. Although this hormone becomes elevated in the milk fever cow, the lag in the bone response prevents rapid enough replenishment of blood calcium.

Alkaline diets impair the bone's response to PTH, but acidic diets (diets to which anionic salts have been added) improve response to PTH.

Active vitamin D has the primary effect of improving calcium absorption from the gut. Higher production of this compound is triggered by low blood calcium, as well as low blood phosphorus. High phosphorus apparently tends to inhibit synthesis. This active vitamin D also is elevated in the milk fever cow, but apparently the lag in response prevents it from doing its job in time.

These findings have led to a couple of conclusions. Moderate intake of calcium (50 to 150 grams per day) during the dry period delays the response to these two compounds and is undesirable. Also, excessive phosphorus feeding (over 50 grams per day) during the dry period could delay the formation of and response to active vitamin D. This would be undesirable as well.

Calcium has to be present in the digestive tract in order for absorption to take place. This means that keeping cows on feed and providing a good calcium intake right after calving is desirable.

Treatment. The method of choice for treating milk fever still remains the intravenous injection of a solution of calcium gluconate. Slow administration is needed to prevent a heart block. Response usually is rapid but relapses are common — about 30 percent of cases. Also, evidence is growing that cows that have had milk fever are more susceptible to other problems such as mastitis (especially coliform), displaced abomasum, retained placenta and ketosis.

Oral gels containing calcium chloride have shown promise in treating cows with early signs of milk fever, and in preventing relapses after treatment with intravenous calcium gluconate. The calcium chloride gels must be carefully administered, however, to avoid injuring the back of the cow's throat, and causing development of a life-threatening abscess there.

The practice of partial milking after calving has been shown to be of no use in preventing milk fever relapses. Because of partial milking's adverse effects on the start of lactation and mastitis flare-ups, it is not recommended.

Prevention. The traditional method of preventing milk fever has been to limit calcium during the dry period. In theory, limiting calcium helps condition the dry cow to calcium deficiency and makes her better able to respond to the high calcium demand that occurs when she begins milking.

Feeding guidelines for dry cows that follow the traditional method of preventing milk fever include:

• Limiting calcium consumption to less than 50 grams per day or feeding calcium at less than 0.5 percent of ration dry matter;

• Limiting phosphorus consumption to less than 45 grams per day or feeding phosphorus at 0.35 percent of ration dry matter;

Meeting these guidelines generally is accomplished by restricting use of high-calcium forage such as alfalfa hay or silage in feeding dry cows.

Replacing some or all of the alfalfa hay or silage with grass hay or silage, small-grain silage, cornstalks or corn silage cuts calcium consumption during the dry period and helps prevent milk fever. This is a common and effective practice. But it can be expensive and difficult since special forages are needed. If calcium intake cannot be reduced to less than 50 grams per day, milk fever may continue to be a problem in the herd.

Recent research shows that supplementing dry cow rations

with anionic salts is another effective method of preventing milk fever. Let's take a closer look.

Anion and cation are terms used to describe minerals in a ration. Whether a mineral element is an anion or cation depends on its electrical charge. Anions have a negative charge, while cations have a positive charge.

Important ration anions include chloride, sulfur and phosphorus. Ration cations include sodium, potassium, calcium and magnesium. Cation-anion difference is calculated by subtracting the equivalent weight (molecular weight divided by the charge) of anions from the equivalent weight of cations.

Rations high in cations relative to anions or having a high cation-anion balance are considered alkaline rations. On the other hand, rations low in cations relative to anions or having a low or negative cation-anion balance are considered acidic.

Alkaline rations tend to cause milk fever, while acidic rations tend to prevent milk fever. The main benefit of feeding acidic rations is increased mobilization of calcium from bone.

Researchers have used anionic salts such as calcium chloride, magnesium sulfate, aluminum sulfate, ammonium chloride and ammonium sulfate to develop acidic (low or negative cation-anion difference) dry cow rations. Ammonium chloride, ammonium sulfate and magnesium sulfate are the anionic salts used most commonly by nutritionists to adjust cation-anion balance.

Reducing milk fever incidence from over 50 percent down to 4 percent or even to zero has been achieved by researchers in the US, Canada and Europe through feeding of acidic rations.

Supplementing of anionic salts during the dry period should begin 3 weeks before calving. The cost is $5 to $8 per cow. Anionic rations also reduce the incidence of retained placenta, udder edema and milk fever, according to recent studies. Other research indicates that cows fed an acidic diet bred back sooner after calving than a control group, and that improvement in milk yield alone potentially can return about $10 for each $1 invested.

Anionic salts can be helpful when there is a high incidence of milk fever or when if is difficult to limit calcium during the dry period or when dry cows are fed high-potassium forages. It's important to have your feed ingredients analyzed, since book values on mineral content can be very misleading.

Anionic salts work best in rations providing high calcium levels during the dry period: over 150 grams per day. **Don't** supplement anionic salts when calcium consumption is low.

Herds in which the late dry cows are component-fed usually receive anionic salts in a pre-mixed form with carriers. An example of a mixture of anionic salts contains 4 ounces of ammonium chloride and 4 ounces of magnesium sulfate blended with 8 ounces of distillers' dried grains. Feed this to dry cows at the rate of 1 pound per cow per day split

Some commercial products offer anionic salts in a pelleted form.

equally between the morning and evening feedings.

Another effective mixture contains 2 ounces of ammonium chloride, 2 ounces of ammonium sulfate, and 4 ounces of magnesium sulfate blended with 8 ounces of distillers' dried grains fed at the rate of 1 pound per cow per day. Bring dry cows up to the full feeding rate of the anionic salts mixture gradually over a three-day period.

Anionic salts are very unpalatable, making it difficult to topdress them in a grain or mineral mix. When supplementing anionic salts, feed the topdress mixture twice per day, and blend with silage whenever possible. Using palatable carriers such as distillers' dried grains and molasses in grain or mineral mixes containing anionic salts also helps.

Storing grain mixes containing anionic salts may reduce the palatability of the grain mix, especially during hot weather. It may be better to mix the anionic salts with the grain just before feeding during summer. Palatability is less concern with a total mixed ration (TMR). Some commercially formulated products are pelleted to reduce separation of feed ingredients and to improve handling.

If late dry cows are fed a TMR, you can add the anionic salts directly to the mixer, without any additional carriers, and feed the TMR just once daily.

More rigid guidelines for adding anionic salts to a TMR are as follows:

1) Analyze forages and concentrates for sodium (Na), potassium (K), chloride (Cl) and sulfur (S).

2) Select feed ingredients with low dietary cation-anion difference (DCAD), especially forages low in potassium (K). The goal in a dry cow ration is to provide an excess of anions, mainly Cl and S, compared to cations, mainly Na and K.

3) Balance ration Mg at .40 percent, dry matter basis. Supplement magnesium sulfate to achieve this.

4) Balance ration sulfur at .40 percent, dry matter basis (supplement with calcium sulfate and/or ammonium sulfate).

5) Add dietary Cl to lower DCAD to -150 meq/kg, dry matter basis, OR a total of <3.0 equivalents/day total anionic salts. Supplement with ammonium chloride or calcium chloride to accomplish this.

Two different calculations will be necessary at this point. First, DCAD is calculated by determining the sum of sodium plus potassium less the sum of chloride plus sulfur. These must be converted from percent to milliequivalents per kilogram (meq/kg) of feed.

Factors to convert from percent to meq/kg are:

Element	Factor
Sodium (Na)	435
Potassium (K)	256
Chloride (Cl)	282
Sulfur (S)	624

After finding the total percent of these elements in the ration, and multiplying percent by the factor to obtain meq/kg, the DCAD is found by this equation:

(Na + K) - (Cl + S)

Here is a sample calculation of the DCAD of a ration:

Percent of elements, dry matter basis: Na = .20%; K = 1.10%; Cl = .85%; S = .40%.

Milliequivalents/kg of elements: Na = 87.0; K = 281.6; Cl = 239.7; S = 249.6.

Dietary cation-anion difference:

(87.0 + 281.6) - (239.7 + 249.6) = -120.7 meq/kg.

Second, the total equivalents of anionic salts must be calculated. The equivalent weights of the anionic salts are as follows: magnesium sulfate, 123 g (4.3 oz.); calcium chloride, 74 g (2.6 oz.); calcium sulfate, 86 g; (3.0 oz.) ammonium chloride, 54 g (1.9 oz.); and ammonium sulfate, 66 g (2.3 oz.).

For example, if 4 oz. of magnesium sulfate and 4 oz. of ammonium chloride are fed, then a total of 3.0 equivalents of anionic salts are being fed (.9 equivalents from the magnesium sulfate and 2.1 equivalents from the ammonium chloride).

If DCAD cannot be lowered to <0 meq/kg with <3.0 equivalents of anionic salts, then consider new feed ingredients with lower DCAD.

6) Check ration nonprotein nitrogen (NPN) and degradable intake protein (DIP); if NPN is > .50 percent, or if DIP is > 75 percent of crude protein, then reduce the amount of ammonium salts added to the diet.

7) Add calcium (Ca) to a daily intake of 150 g and add phosphorus (P) to a daily intake of 45 g. (Sources: limestone; monocalcium phosphate; dicalcium phosphate; monosodium phosphate).

8) Monitor dry matter intake. You may have to reduce the amount of salts if they impair intake until it improves, or use additives/feeding methods to make feed more palatable.

Though these recommendations will often work, at times they won't. Remember that anionic salts will not prevent problems at calving caused by overconditioning or feeding dry cows rations that have not been balanced properly for fiber, energy and protein. When you supplement anionic salts, it is especially important to be accurate both in mixing and delivering the ration to the cows. Supplement anionic salts only after consulting with your veterinarian and nutritionist.

Ketosis

Occurrence. Ketosis usually occurs 10 days to 6 weeks after calving in high-producing cows. Peak incidence is about 3 weeks after freshening. In a 1990 survey, Hoard's Dairyman readers reported over 50 percent of their herds had the problem involving about 5 percent of the cows.

Many high-producing cows go through a borderline type of ketosis in early lactation when milk output exceeds nutrient intake and body reserves must be used. So, the incidence reported depends on how hard you look for it.

Symptoms. First there is a lack of appetite, especially for grain. Rumen inactivity and dry feces occur in some cases. Usually the cow is dull and listless, but occasionally there is the "nervous" type in which the cow is highly excitable. There is a gaunt, dull appearance, loss of weight and lowered milk production. Sometimes there is incoordination, due to general weakness. Death is rare.

When a cow has other problems such as retained placenta, hardware or displaced abomasum, these conditions often predispose her to ketosis. This then is called secondary ketosis, and represents about one-third of all cases. Obviously, the complicating problem has to be cleared up before there will be response to ketosis treatment. Temperature is not elevated in true or primary ketosis, so an elevated temperature suggests complications.

In ketosis, two major changes occur in the blood. One is a drop in sugar or glucose, which is the initiating factor. The second is a rise in ketones, which are compounds produced in the liver from mobilized fat. The name ketosis comes from the elevated ketones. One of these is acetone, hence the earlier name, acetonemia.

Elevated ketone levels in the blood are reflected in higher levels in urine and milk, either of which can be used to check for ketosis. Milk has about half the ketone level of blood, while urine has about four times more. The usual rapid test is to place some urine or milk on a little mound of test powder on a white card. A positive test is a pink to purple color. Commercially available dipsticks test urine (but not milk).

Since urine has about eight times more ketones than milk, it is very sensitive and becomes positive before milk. Many high producing cows in early lactation will have a positive urine test but will not necessarily require treatment. A negative urine test rules out ketosis, but a positive one is more difficult to interpret.

The milk test is a more conservative but more accurate indication of the need to take action. Check with your veterinarian about the test powder. When the powder becomes pink with milk, there is at least a borderline problem, and treatment should be initiated.

Cause. The general consensus is that ketosis is due to the imbalance between outgo of milk and intake of nutrients. The cow in early lactation has a critical need for nutrients she can make into blood sugar. A cow milking 100 pounds per day needs about 5 pounds of glucose from the blood to make

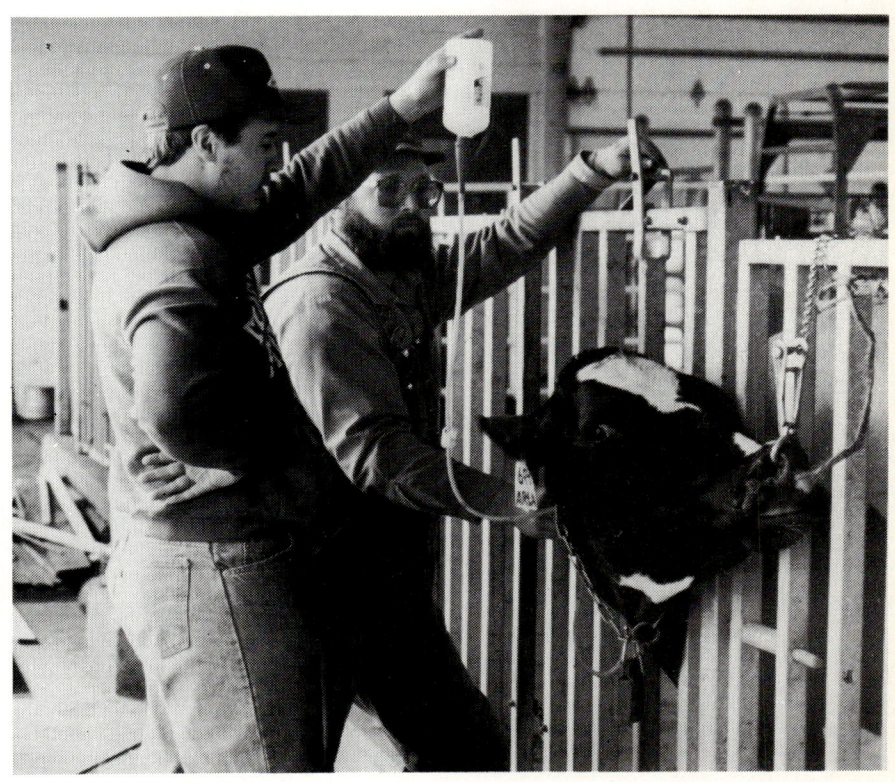

the sugar in that amount of milk.

When blood sugar is too low, the body tends to compensate by mobilizing fat from body stores. When this fat gets to the liver, some is converted to ketones and some is deposited in the liver. The latter is especially true with prolonged ketosis. The fat deposited in the liver makes it even more difficult for the liver to make blood sugar. This causes still more ketones to be formed. The symptoms of ketosis are probably due to the toxic effect of high blood ketone levels.

Treatment. Most of the accepted treatments attempt to raise blood glucose in some way. Here are the most common ones:

Intravenous glucose injections. Usually, 500cc of a 50 percent dextrose solution is used. This is the most rapid and direct way to supply sugar, but much of it spills over into the urine and the rest is used up in a few hours. So relapses are common when this is the sole treatment. Often, glucose injection is followed up by one of the longer acting treatments.

Hormone treatment. Although there actually is not a hormone deficiency, hormones called glucocorticoids (cortisone) are useful for treatment. These stimulate the body to produce more blood sugar by forming it from body protein.

Prolonged use of glucocorticoids is not desirable because they deplete body protein and reduce disease resistance. The latter is particularly important if the cow also is fighting infectious diseases such as mastitis or metritis. In these cases, antibiotics may be administered along with the steroids.

A hormone called ACTH, which stimulates the body's adrenal gland to produce more of its own glucocorticoids, sometimes is used.

Oral sugar precursors. Propylene glycol or sodium propionate are two compounds commonly fed or drenched. The cow uses these to produce glucose in the liver. Propylene glycol has become the first choice because of its palatability, ease of handling, low cost and absence of sodium.

Propylene glycol also has the advantage that an outside source of glucose can be supplied in a gradual and continuous fashion until the cow recovers. It can be fed in the early stages when the cow still is eating. It also can be given by drench, but make sure it doesn't go into the cow's lungs. Often it is used as a follow-up of glucose or hormone treatments. Usual dosage is 8 to 16 ounces per day, with the higher level being given in split dosages. Prolonged, high doses of propylene glycol may adversely affect rumen bacteria.

Miscellaneous treatments. Insulin may be administered by a veterinarian to speed breakdown of sugars to glucose. Chloral hydrate, a sedative, also raises blood glucose and can be useful in nervous cases. Cobalt sometimes is given if a deficiency is suspected.

Prevention. While there is no known procedure that will guarantee a cow won't have ketosis, these suggestions should help:

1. Prevent other diseases which may lead to secondary ketosis.

2. Have cows in the proper body condition at calving with a body condition score of 3 to 3.5 on a 5-point scale. Proper body condition is a tradeoff between the energy stores necessary to allow for high milk production during a period of negative energy balance and the appetite-diminishing effects of excessive fat.

3. Feed 5 to 10 pounds of grain per head per day for about 3 weeks before calving. This practice requires that close-up dry cows be handled separately. Owners of smaller herds may be reluctant to divide dry cows into two dry groups. However, the eventual savings from preventing problems after freshening will more than justify the cost and the trouble.

4. Encourage maximum energy and dry matter intake after calving. Feed high-quality forages. Forage intake is closely related to its fiber content; neutral detergent fiber appears to be the best predictor of animal intake. Feed total mixed rations if possible. If not, then limit grain to 6 to 8 pounds per feeding.

5. Feed grain mixes that have relatively high fiber concentrations. This can be accomplished by using ground ear corn or by adding beet pulp, whole cottonseed, dry brewers' grains or similar feeds to ground shell corn. This becomes more important with lower quality forages.

6. Feed total rations with 18 to 19 percent crude protein to cows in early lactation.

7. Avoid rations with more than 50 percent moisture (less than 50 percent dry matter).

8. Feed 6 grams niacin per head, starting 3 weeks prior to calving, and continue through the first 100 days of lactation in herds with ketosis problems.

In more severe problem herds, use a milk test weekly on fresh cows, and feed or drench propylene glycol at a level of 4 to 8 ounces daily to positive cows.

Fat cow syndrome

Occurrence. In recent years the term fat cow syndrome has been used to describe a condition occurring within a few days of calving among cows that are excessively fat. Often these are cows with breeding problems in the previous lactation that have had very long dry periods.

Symptoms. The condition is characterized by depression, lack of appetite, suppressed immune system to combat infections and general weakness. It is referred to as a syndrome because it almost always is associated with other problems at calving, such as milk fever, displaced abomasum, retained placenta, metritis or mastitis. Often, there is an elevated temperature due to an associated infection.

Although blood ketone and fatty acid levels usually are high, the ketosis almost always is secondary to another problem. Blood glucose may be high or low. These cows often die and have fatty livers and a lot of internal fat. Since a number of clinical signs may be present in this syndrome, it's important for your veterinarian to establish a diagnosis. The diagnosis often is confirmed by unfavorable response to conventional treatments for ketosis.

Cause. Fat cows are predisposed to develop fatty liver before and after calving. Though most fat cows have fatty livers, all fatty liver cows are not fat cows. Many times in chronic ketosis, a thin cow can develop fatty liver because she's mobilizing what little fat she has.

It is important to understand how a fatty liver develops. The newly fresh cow typically doesn't consume enough energy after calving to satisfy her energy needs. So she pulls the fatty acids from her body fat stores. The cow uses fatty acids for milk fat synthesis, but her liver takes in a majority. The more fatty acids are mobilized, the more the liver absorbs.

From these fatty acids, the liver manufactures another fat called triglyceride. Ideally, the cow should take up the triglycerides for milk fat synthesis, but in ruminants, this is unfortunately an inefficient process. Fatty liver develops because the liver manufactures triglycerides faster than triglycerides move to the udder.

Once triglycerides accumulate in the liver, it is very difficult for them to be removed. These triglycerides then interfere with important liver cell activities such as producing immune cells, neutralizing toxins absorbed from the intestinal tract, and creating blood sugar from other compounds. This explains why fat-cow syndrome cows often die.

Treatment. Treatment is not very effective. It usually consists of intravenous glucose, with antibiotics to combat infections. Excellent care at calving is critical. If excessively fat cows can get through the calving period without complications, they may be able to adjust to the mobilization of large amounts of fat. But they still are more susceptible to problems and are less likely to get off to a good start. In many herds, it may be most economical to cull extremely fat cows before they calve.

Prevention. The key is to avoid excessive fatty acid mobilization from body fat. The cow's body condition is critical, and this is best attended to long before she is dried off.

Fat cows have a sluggish appetite that stimulates excessive fat mobilization. But it's risky to attempt correcting this by restricting feed during the dry period, since this will stimulate more fat mobilization. The best time to adjust body condition is during lactation. Work with a nutritionist so you can avoid modifying a cow's body condition during her dry period.

If a cow does enter the dry period with too much body condition, it is especially important that the far-off and close-up dry cow rations be very well formulated. The early lactation ration also must be very carefully managed. Do not increase grain feeding by more than 1/2 pound per day after calving. More rapid introduction of grain will cause rumen acidosis, which will make the fatty liver even worse. Provide exercise for dry cows. And finally, maintain a 12- to 13-month calving interval to avoid long dry periods.

Retained placenta

Retained placenta or failure to "clean" after calving occurs when the placenta (tissue surrounding the fetus and lining the uterus during pregnancy) fails to separate from the uterine wall. The attachment sites are called cotyledons or "buttons" and are the places where nutrients from the mother's blood pass to the calf during fetal growth.

Normally, a cow should clean within an hour or so after calving. If she does not clean within 12 hours, you can consider the placenta retained.

In cows that simply retain their placenta and do not develop an infection, fertility is normal. However, infections of the reproductive tract that frequently accompany retained placenta are more serious. These infections will mean a slower recovery of the reproductive tract after calving, longer intervals to first estrus and first breeding, lower conception rates, longer calving intervals and culling more cows due to reproductive health problems.

Occurrence. Most surveys, including those of Hoard's Dairyman, report an incidence of 5 to 10 percent. Retained placentas are more common in first-calf heifers and in older cows.

In heifers, it frequently is associated with difficulty in calving. In older cows, it may be due to poor muscular contractions and accumulated uterine health problems from earlier calvings.

Incidence is higher in cows that retained at the previous calving or had a longer calving interval, and incidence is very high in cows that calve before their expected due date.

Retained placenta also is more frequent in high-producing cows, cows that give birth to twins and in Holsteins compared to Jerseys. It is believed to be somewhat heritable.

Symptoms. The symptoms are rather obvious. The placenta is not found in the calving area, and usually a string of cleanings is hanging from the vulva. Within a day or so, a characteristic bad smell develops.

On occasion cows try to eat the cleanings, and there is danger of choking. Remove the placenta from the calving pen as soon as it is found.

Cause. The causes of retained placenta can be grouped into two categories: interference of the process that loosens the connections between the placental attachments and the uterus; lack of or weakened uterine contractions.

The loosening of the placenta is a series of physiological events. If one is disrupted, the placenta will be retained. Placentas mature in the last months of pregnancy, and hormones influence changes in the placental lining and the caruncles of the uterus. The final maturing process depends on an estrogen hormone for at least five days before the due date. Therefore, when cows calve five or more days before the due date, they frequently will retain their placenta.

Certain pressures within the uterus caused by uterine contractions are a physical aid to placental removal. Microscopic, finger-like projections (villi) form the attachments within the placentomes (the combined placental cotyledon: uterine caruncle union). Contractions cause both anemia and hyperemia (swelling) within the villi. The anemia makes the lining of the villi shrink. Contractions also shrink the size of the uterus and the caruncles, making expulsion of the placenta easier. Anything that stops or weakens muscle contractions, such as milk fever, will interfere with expulsion of the placenta.

Any disease or infective organism that causes an infection in the reproductive tract, produces a high fever or contributes to abortions or stillbirths will raise the incidence of retained placenta. Common disease problems include brucellosis, bovine virus diarrhea (BVD), infectious bovine rhinotracheitis (IBR) and leptospirosis.

Severe deficiencies of vitamin A or beta-carotene (converted to vitamin A by the cow), selenium, iodine and improper levels of calcium and phosphorus in the diet (which also can contribute to milk fever), all can boost incidence of retained placentas.

Treatment. Treatment is a job for your veterinarian. In most cases, it involves injecting antibiotics in the early stages, usually intramuscularly. Be sure to observe the recommended milk withholding periods for the antibiotic involved.

If a long string of cleanings is hanging from the vulva, most veterinarians suggest cutting it off, leaving a foot or more hanging out. A longer amount is unsightly and serves as a wick to carry infection back into the uterus.

There are differences of opinion regarding the advisability of cleaning the cow manually. In recent years, the trend has been to use antibiotics and let the cow clean on her own. But this is a decision for your veterinarian based on each specific case.

Most veterinarians wait at least 1 to 3 days before cleaning a cow manually. Of course, the danger is possible damage to the uterus if the placenta is not ready to come away.

Antibiotic therapy in the uterus, as well as parentally, is of value to prevent infection.

Prevention. Overconditioning of cows during late lactation and dry periods results in more retained placentas. Because of the many nutritional factors involved, well-balanced and properly delivered dry cow rations are a must. Since elements like selenium are toxic in excess, do not add them to the ration indiscriminately.

Effective prevention of milk fever also helps prevent retained placenta. Calcium given intravenously or orally after calving will reduce retained placentas. The calcium corrects poor muscle tone in the uterus.

Oxytocin or prostaglandins given within 12 hours after calving may be of some value. However, just letting the calf nurse 3 to 4 hours after calving stimulates oxytocin release in the new mother, and can help her clean. Of course, the calf should receive its first feeding of colostrum well before this, as soon after birth as possible.

Maintain a vaccination program against brucellosis, leptospirosis, IBR, PI-3 and *H. somnus.*

Don't allow dry cows to become stressed due to heat, humidity, poor ventilation or crowding. Stress near the time of calving may cause cortisol release, which causes the cow to calve prematurely. Mastitis during the dry period also may cause premature calving, so maintain an effective dry cow mastitis program. Avoid moving cows which have just calved. Some dairymen report success with providing warm water and beet pulp after calving.

Even if your incidence of retained placentas is within acceptable limits, keep uterine infections at a minimum by providing good sanitation at calving. Provide a dry and clean calving pen, and sanitize hands and equipment when pulling calves. For bedding, dry straw is preferable to sawdust.

Displaced abomasum

Displaced abomasum, also called a DA, twisted stomach or twisted gut, is a condition in which the fourth or true stomach of the cow's four stomachs is twisted to the left or right from its normal position. Feed passes first to the rumen, reticulum (where hardware accumulates) and omasum before reaching the abomasum. By this time the particles are small and the fiber has been broken down.

The normal position of the abomasum is near the belly floor on the right side. The large rumen and reticulum are located on the left side and extend from the belly floor up the left side. The omasum is on the right side above the abomasum. (See diagram.)

About 80 to 90 percent of the time when a displacement occurs, it is to the left. The abomasum migrates from the right side of the belly under rumen and up on the left body wall. There it is trapped or squeezed between the rumen and the left side of the cow.

Although there may be a twist or turn, it is most common for the abomasum to be in the normal, upright position. However, the entrance and exit of the abomasum are restricted because of the pressure exerted on them from stretching the abomasum around to the other side of the rumen. The abomasum becomes filled with gas and bloats. The veterinarian taps the cow's left side and listens for the "ping" associated with this bloating.

In a right side displacement, the abomasum moves up the right side and may twist from front to back, forming a torsion. The abomasum is trapped with the front tipped back or rotated above the omasum. The abomasum becomes filled with fluid and some gas, but usually is not bloated or distended with gas. The right displacement normally is more severe, and recovery rate is much poorer compared to a left displacement.

Occurrence. About 90 percent of displacements occur within 6 weeks of calving. In a recent Hoard's Dairyman survey, 51 percent of those responding reported having had the problem, with an incidence of over 3 percent of the cows.

Symptoms. As might be expected, if the digestive tract is restricted or blocked, the cow goes off feed and production drops. She may stand with an arched back. There is very little manure at first or mild diarrhea, followed by very dark, bad smelling feces or diarrhea.

Temperature is normal unless there is an infection. Generally, the cow is depressed and dehydrated. Some of the symptoms resemble ketosis. The urine or milk test for ketosis may be positive, but the cow will not respond to ketosis treatments.

Cause. The exact causes of displaced abomasum are not completely understood, and they vary from case to case.

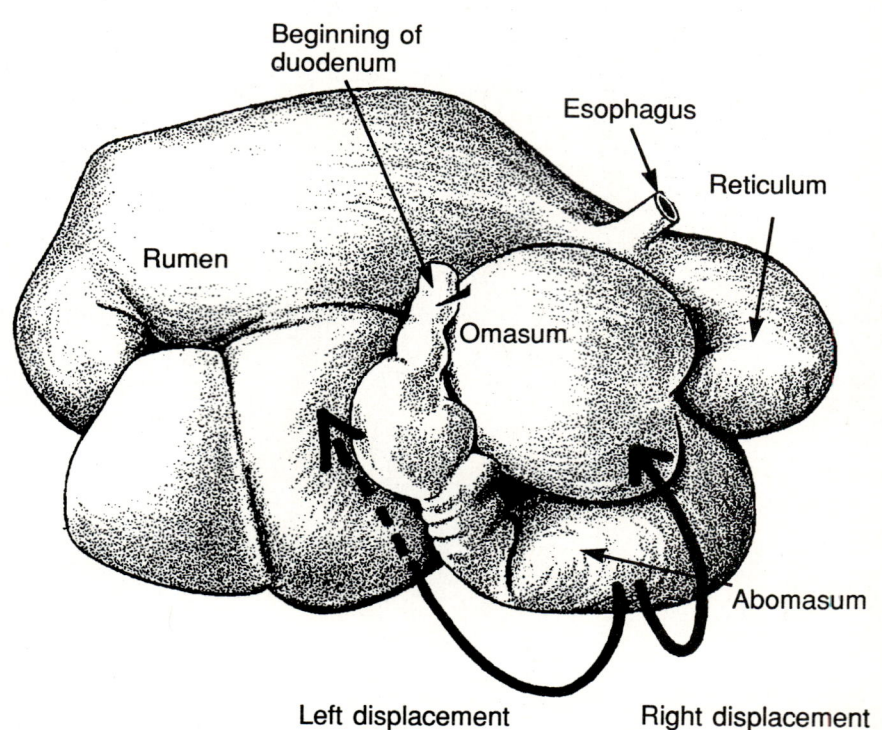

Des Moines Skyline

State Capitol Building, Des Moines

Storm Lake

County Courthouse, Dubuque

Sailing on Grey's Lake

Cedar Rapids

Vander Veer Park, Davenport

The Old State Capitol Building, Iowa City

Spirit Lake

Herbert Hoover Birthplace, West Branch

Famous "Little Brown Church in the Vale," Nashua

Bellevue, on the Mississippi River

Mississippi River at Burlington

Ottumwa

Graceland College, Lamoni

Lidke Grist Mill, North of Decorah

Lake Okoboji

Pella

University of Northern Iowa, Cedar Falls

Governor's Mansion, Des Moines

John Wayne's Home, Winterset

Locks on the Mississippi River at Keokuk

Salisbury House, Des Moines

Mason City

Kalona Historical Village

View of Mississippi River from Eagle Point Park, Dubuque

Drake University, Des Moines

Iowa State University, Ames

McGregor, on the Mississippi

Herbert Hoover Presidential Library, West Branch

Missouri River, North of Council Bluffs

Near Mount Vernon

Historic "Goldenrod," Birthplace of 4H, Clarinda

Farming near Pella

Wild Cat Den State Park

Northeast Iowa Farmland

Grist Mill at Wild Cat Den State Park, Muscatine

Fort Atkinson

Farming near Maquoketa

Bellevue